RECOLLECTING THE FIRST VISION

Copyright © 2021

Gospel Tangents

All Rights Reserved

Except for book reviews, no content may be reproduced without written permission.

(Note this conversation was recorded on Jan 20, 2021. The interview has been lightly edited for clarity.)

Contents

Introduction 3
Memory Problems with First Vision 4
Were Revivals in Palmyra in 1820? 15
Did Methodist Minister Scold Young Joseph? 25
Comparing the Primary Accounts 31
Orson Pratt's First Vision Influence 38
Additional Resources: 43
    Dan Vogel on First Vision 43
Last Thoughts 44

Introduction

The First Vision is the founding event of Mormonism. Critics of Joseph Smith have questioned Joseph Smith's accounts of First Vision. Were there revivals in 1820 near Palmyra? Did Joseph lie about seeing God? Did a Methodist minister really scold Joseph Smith? BYU professor Dr. Steven Harper looks at the malleability of memory. Can Joseph's account be trusted? Are there different ways to explain discrepancies in Joseph's various accounts of the First Vision? Dr. Harper will answer these questions as we discuss his book "First Vision: Memory and Mormon Origins." We will also discuss Orson Pratt's outsized influence on helping the entire Church remember Joseph Smith's First Vision. Check out our conversation...

Tags: Gospel Tangents, Rick Bennett, LDS Church, Latter-day Saints, LDS Church, Mormon, Mormon Church, Church of Jesus Christ of Latter-day Saints, Mormon history, Mormon, LDS Church, LDS, Church of Latter Day Saints, First Vision, Orson Pratt, Joseph Smith, Steven Harper, memory, memory problems, Dan Vogel, Sandra Tanner, memory, are memories accurate, anachronisms of First Vision, Orson Pratt, Wesley Walters, Joseph, vision, Methodist minister, memory, Methodist, accounts, Milton Backman, Palmyra, memories

Memory Problems with First Vision

Introduction

The First Vision is the founding event of Mormonism. Yet, critics of Joseph Smith's First Vision account claim that Joseph changed his accounts over the years, resulting in contradictions between the four primary accounts. Could there be other ways to explain these discrepancies? Dr. Steven Harper from BYU has written a book called "First Vision: Memory and Mormon Origins"[1] that seeks to answer some of these questions. Check out our conversation....

Interview

GT 00:02 Welcome to *Gospel Tangents*. I'm excited to have an awesome expert on the First Vision. Could you go ahead and tell us who you are?

Steven 00:11 Sure. My name is Steve Harper, and I'm "no one of consequence, "to use that line from *The Princess Bride*. I teach Church History and the Foundations of the Restoration at BYU. I worked on the *Saints*, four volume series, with a bunch of other folks. I've been a volume editor of the *Joseph Smith Papers* in the past, and so I studied the early history in the revelations of Joseph Smith.

GT 00:43 Great. Well, why don't you tell us a little bit about your academic background? Where did you get your bachelor's and master's and Ph.D. and all that stuff?

Steven 00:50 I went to BYU as an undergraduate. I majored in history after dipping my toe into ancient Near Eastern and Biblical

---

[1] Can be purchased at https://amzn.to/39XqDQY

Studies and getting out of that when I couldn't hack it. I went from there to Utah State for a master's degree in American History. I had studied paleography, early American handwriting, as an undergraduate and had some really formative experience working with early Latter-day Saint documents, revelation manuscripts, journals, especially William McLellin's journals and revelation manuscripts.

GT 01:30 Did you read any Hofmann forgeries while you were at it?

Steven 01:33 I did, even as a teenager, I was reading Hofmann forgeries, unbeknownst to me, until sadly, he killed a couple of people, as you know, and the whole thing unraveled on him.

GT 01:46 He was coming up with something with William McLellin.

Steven 01:48 Exactly right. That's why, in fact, as an undergraduate, the opportunity came to work on McLellin's documents, because he had rumored that he had McLellin's Collection, or he could obtain it. When the Church discovered that it had it all along, it was eager to get it published. The rumors were that the Church was suppressing documents. Hofmann was the one who actually started those rumors. So, the Church was eager to publish the McLellin Collection. They asked Jan Shipps, a very, very excellent scholar of the Saints and American religion generally, to be the editor. She collaborated with John Welch at *BYU Studies*. So, *BYU Studies* and the University of Illinois Press co-published the journals of William McLellin. I was an editorial assistant at *BYU Studies*, and I got to read them very closely with Jan Shipps. It was a highlight of my undergraduate career to work closely with Jack Welch and Jan Shipps to read those documents and learn from them how to do very close readings of source material and documentary editing. I was hooked. I went from there to Utah State and did a master's thesis on what early Latter-day Saint missionaries taught, how they taught, who they reached socioeconomically speaking, what the determinants of conversion were in that environment, and I've been working on those kinds of questions ever since.

GT 03:24 Wow, that's awesome. So, your undergrad was at BYU. Your master's degree was at Utah State.

Steven 03:32 Then, I didn't know what to do. I wanted to be a professor. I wanted to teach the History of the Church. I wanted to teach Joseph Smith's revelations. So, I tried to figure out the way to do that. My early Church History professor had said, "Go get a Ph.D." So, I was on that track, but [I didn't know] how to pay the bills in the meantime, etc. So, I took a job microfilming valuable genealogical records for the Church's Family History Department. That took me to Richmond, Virginia. That was a great experience.

GT 04:10 Was that the Wahoos?

Steven 04:12 Yeah, I think so.

GT 04:13 I have a Virginia shirt. I've been to a football game there.

Steven 04:16 I think they're in in Charlottesville.

GT 04:19 Oh, that's right.

Steven 04:20 That's 90 miles away, the University of Virginia.

GT 04:23 Who's in Richmond?

Steven 04:23 The University of Richmond.

GT 04:24 Oh, okay.

Steven 04:25 And VCU.[2]

---

[2] Virginia Commonwealth University.

GT 04:26 Oh, they're the [Richmond] Spiders.

Steven 04:27 Yes, exactly. So, it was a great experience. I met some really wonderful people there and had some great experiences. I remember filming from as early as I could get into the library, right up until lunch hour and then spending my lunch hour in the reading room, digesting anything they had on American religion and so forth. I worked on an article there and published an article that won an award. It won the Best Article of the Year Award from the Mormon History Association during that year. That was a good experience for me. I went from there to Alexandria, Virginia. Well, I actually worked in Washington, D.C., filming records. It sounds sort of sexy, but I was in the D.C. Records Center and a very sort of sketchy place in Washington D.C.

GT 05:30 You didn't run into Nicolas Cage looking for hidden treasure or anything? (Chuckling)

Steven 05:33 I didn't, that would have been cool. But I didn't. I went from there to Bethlehem, Pennsylvania, where I continued filming records, but also started a Ph. D. program at Lehigh University, which was a great fit for me. It was a small, little school, with Ivy League envy, but a really good early American...

GT 05:42 They spell Lehi wrong, though.

Steven 05:57 They do.

GT 05:58 They stick a 'gh' on the end.

Steven 05:59 They haven't, I don't think, seen the Book of Mormon version of it, or the Roller Mill.

GT 06:04 (Chuckling)

Steven 06:06 So, I'm not sure they know how to spell it. But I enjoyed that experience very much. It was a good fit for me.

GT 06:14 So, you got your Ph.D. from Lehigh?

Steven 06:16 I did.

GT 06:17 Oh, wow, very cool. I see we're in a room and you can kind of see behind me. You've got a lot of sports memorabilia. I don't talk to many historians who are as big of sports fans as me. I think you might be.

Steven 06:30 Well, before I ever imagined a person could make a living being a historian, I thought the only option was to be an NFL quarterback or a first baseman, maybe in Major League Baseball. If I had had the ability to do that, I maybe would have gone on to do that.

GT 06:50 I would have, too. I was going to play shortstop for the Dodgers.

Steven 06:54 Right. That would be a good choice. I enjoy it, and I've got jerseys from my kids and my father-in-law's letterman sweater and one of his track shoes. He holds a record for the hurdles in his southern California high school. It's a fun room where we collect stuff from the kid's accomplishments.

GT 07:15 Well, I know people can't see behind me, but there's a Blackfoot basketball [framed jersey] right above your head there. My mom was from Blackfoot, but there was no Blackfoot High School when she was around.

Steven 07:27 That's a practice jersey from being a benchwarmer on the high school [basketball team].

GT 07:32 That's what I was too, a benchwarmer.

Steven 07:35 That's all that is.

GT 07:36 Although, I can say that we did take state, so at least I was a benchwarmer for the state champs.

Steven 07:40 That's impressive.

GT 07:43 All right. Well, cool. It's sad to me that people don't like my little jabber about sports. I'm a big sports fan. It's nice to find a fellow fan. That's great.

GT: Well, tell us about your book. You've got a book on the First Vision, go ahead and hold it up, a little higher. So, this is your latest book on the First Vision: Memory and Mormon Origins. So, tell us about your book.

Steven 08:13 Well, it is an ambitious work. I like to always be doing something that really pushes me, that stretches me, and I wanted to publish something with an academic publisher, a prestigious academic press. Oxford was as high as I could aim. A lot of other great books were coming from them on Mormon studies topics and all kinds of topics. So, I pitched an idea to them that would be the first marriage between Mormon History and Memory Studies. Both of those fields are robust. There's lots of interesting advances, really great things being published. But nobody had ever done a full-length study that really welded those two kinds of things together. So, I wanted to dig into the historical record of the First Vision, with the methodology and interpretive framework that was based a lot on memory studies, the science, the neuroscience and the psychology of memory.

Steven 09:30 There's a theoretical book by Thomas Anastasio and his colleagues, about the consolidation of memory, how short term and especially long term memories consolidate, and how that happens inside an individual and on a collective basis, and how those processes are analogous with each other. That book was compelling to me. I was quite taken by its premises, by its findings, by its models, theoretical capabilities. I found that the historical record of the First Vision mapped onto that very nicely. So, I used that theoretical framework and a lot of other psychology of memory and intimate study of the First Vision sources, the record that Joseph Smith and others left us, and that was the groundwork for what I thought was a great idea. Other people seem to buy into that. Oxford University Press bought into it. The timing was good. I ended up ultimately getting it ready for the Bicentennial. We published the book in 2019, and, of course, 2020 celebrated the Bicentennial of the First Vision. Timing was good. That gave me plenty of time to work on it.

Steven 10:59 I decided that the first part of the book would be about Joseph Smith's memory. How did he consolidate? How did he form his memories of the First Vision? And what's the nature of those memories? The premise, really, from the beginning is that memories are not what we often think they are. Memories aren't like some kind of data that you can record to a DVD or keep in a file folder somewhere where you just retrieve it, and it's the same data every time. We talk about memory that way, but that is not how it works. An autobiographical memory, like Joseph's memories of his First Vision, are real-time creations. He produced the memory every time he told the memory or recorded it. He produced it out of the past, for sure. There were components of memory that he used, but he also always had some present context that was very essential to the way the memory was shaped. This is what we all do. We might think we don't. But this is what we all do. We have a present situation. It prompts us in some way or other to think about our past. We gather up pieces of the past, and we fuse them together, and form them in a way that makes sense in our present, and that addresses the needs of our present. That's how Joseph Smith came up with his memories of the First Vision. So, people might ask, are they accurate or inaccurate? It depends on what you mean.

Steven 12:41 Memories are accurate, and memories are inaccurate, both. They're not perfect or distorted. They're both of those things. There's no way around that, not for anybody. So, memories are what they are. His [Joseph Smith's memories] are fascinating. The first chunk of the book tells about how he formed those memories. What was the present context, in which he formed each of his various memories of his First Vision that we have record of? Then the second part of the book is how a collective memory first formed. How did the earliest Latter-day Saints besides Joseph Smith, who came to know about his vision, remember it? What roles did they have? How did that work? That part goes up through the canonization of Joseph Smith's history in the Pearl of Great Price in 1880. Then, the third chunk of the book is about contested memory. It's about the fight over what the First Vision means. [The fight] over whether Joseph's memories are accurate or inaccurate or distorted or made up or half-remember dream, as Fawn Brodie said, or all the various claims. The stakes have really been raised on the First Vision in the last 50 years or more, and so that's a compelling story.

GT 14:07 Yeah, definitely. I'm curious, because you've got a background in history, and you're going into all this memory stuff, which sounds more like a scientific thing. Did you consult with a neurologist or a memory expert, or something, as you wrote this book?

Steven 14:21 I did. When I first started, I talked a lot to my brother, who's a Ph.D. psychologist. I said, "What about this idea? What about that?" He said, "That would work. That'd be good." He pointed me in the right directions. "Here are some things you need to read. Here are some things you need to stop assuming." One thing that historians commonly assume is that memories are like something that you can carbon date, that there's kind of a predictable rate of radioactive decay attached to a memory. You'll hear people talk about it like that. "Well, this memory was 18 years after the vision, so it's not as good as one that was 10 years or 12 years after the vision." There's no basis for making that judgment.

It's an assumption. But there's no good criteria that's testable or verifiable. It's unscientific, in other words.

GT 15:24 Doesn't that fly against the normal training of a historian, though, because usually you say the first accounts are the best accounts, and then they get worse as time goes on.

Steven 15:34 That's my point. That's the assumption of a historical method. On what is it based?

GT 15:44 I mean, don't we have centuries of historians that do it that way?

Steven 15:48 Maybe so, but a point I want to make in the book is, it's much better to take each memory on its own merits, evaluated itself. What is the present? What is the present set of circumstances that surround it? And how has it been transmitted down to us? That has a lot more to do with how much confidence we can place in it, than if it was immediately after the vision or much later. It's not as clear-cut as, "recent memory is good and a distant memory is bad." There is a whole lot more potential for interpretive memory, in a memory that's many years removed. Let me say what I mean about that.

Steven 16:40 There's factual memory, right? If you're thinking in terms of Joseph Smith's vision accounts, the factual memory is things like, "I read James 1:5. I went to the woods. I said a prayer after I knelt. I saw two percentages. A few days after, I talked to a Methodist minister. He said these things." [This is] factual memory. Those are the kinds of things that you could think of as sort of objectively true, that somebody outside of Joseph Smith's mind could have observed, perhaps. But there's a whole lot of subjective experience that you can only capture from Joseph's mind. That also changes over time, and that gives us interpretive memory, which is fascinating. It's at least as interesting as the facts of the matter.

Steven 17:38 So, I sometimes say it this way: When you've got Joseph's accounts spread out from 1830 to 1842, you can experience the vision, not just as he experienced it at the time, you can tell some about what he experienced over time. One good way to see that is to look closely at the accounts as they interpret the vision. There are 600 words, or so, of interpretive memory, in the account that's in the Pearl of Great Price. "It felt like I was Paul before Agrippa... It seemed like ever since I was an infant, they persecuted me... I have reflected then and since..." All those things tell us that Joseph is doing work in his mind about what this feels like, over time. He's reflecting on the vision, and as he does so, he finds meaning in it that wasn't there the day of. He couldn't have walked out of the woods and had that interpretation. He would have had different interpretive components to work with in the short term, than he has 12 years later, 18 years later.

Steven 18:53 One of the most interesting things is to not reduce memories to thinking that they're reliable, to the degree that they're proximate to the event; and unreliable in proportion to how distant they are from the event. That's not a good way to gauge it. There's no way to test it. It's just an assumption. So, I've tried to abandon that assumption and evaluate every one of the accounts on a much more scientifically sound criterion. How does Joseph remember based on what his present is, and what his past was? That's the better way, I think, to evaluate a memory.

GT 19:41 I know, I interviewed Dan Vogel a couple years ago.[3]

Steven 19:44 You might have heard me paraphrasing Dan Vogel a minute ago, when I said, "How did he remember the vision at the time? How certain can we be, given that the records are distant," Dan Vogel asks, "that Joseph remembered the vision as he experienced it at the time?" My answer to that is, I don't know. But one thing we can say is we can watch him remember it over time.

---

[3] See our interviews at https://gospeltangents.com/category/dan-vogel/

GT 20:08 Yeah. Well, it was interesting to see how you--like you said, in part one, you started with Joseph Smith and his memories. Then in part two, you talked kind of about the collective memory of the Church. I know you really got into Orson Pratt. It seems like he was a very big driver of how all of us remember, because of a lot of his sermons in General Conference. Then, the end was that kind of contested area.

Were Revivals in Palmyra in 1820?

Introduction

Wesley Walters was one of the first people to question Joseph's Smith's account of the First Vision, saying there were no reports of revivals near Palmyra, NY in 1820 as Joseph Smith claimed. Is there another way to interpret this? BYU Professor Steven Harper is the author of "[First Vision: Memory and Mormon Origins](#)" and seeks to answer this issue. Check out our conversation....

Interview

GT: So, I guess one of the things that I found really striking was why you chose to start with the official version first, the 1838 version, the version we have in the in the Pearl of Great Price. Why did you choose that one, instead of the 1832 version?

Steven 20:52 Great question. I picked that one because the catalyst that caused Joseph to remember, as he did, I think the most important determinant of that, is the rejection he received from the Methodist minister. So, the 1838 account is the one that tells us that. "A few days after the vision, I happened to be in company with a Methodist minister" who'd been influential in getting up the religious excitement. Joseph anticipates a welcome response, a validation of his experience. He tells the minister, and he gets, instead, an unexpected and traumatic rejection. James Allen, the founding father of First Vision studies, was the first to suggest that that event, that rejection, shaped the way Joseph thought about and remembered and told about his vision. I believe that. The more I thought about memory, and Joseph's memories, in terms of psychology, the more I thought that Professor Allen's thesis was valid, that there was really something to it. So, I interpreted Joseph's tellings of his vision, all of them, as, in one way or another, responses to that first experience of telling the vision. So, that's why I started with 1838.

Steven 22:33 Then I backtracked and said, in 1832, he is trying to satisfy that minister. He's trying to tell the experience in a way that is acceptable to the culture, to the Protestant/evangelical culture. He's trying to make it sound like a generic enough conversion account that he's not going to get in trouble for it. It's not going to be rejected. Nobody likes to be rejected. It was traumatic for Joseph to be rejected. I'm not suggesting that he consciously thinks all these things. I don't see any evidence that he deliberately decides to remember in these ways. I think, rather, these are unconscious kinds of decisions. I think that ever after the minister rejects him, he has to respond to that experience in some way or other. When he tells the vision, part of what he's doing is responding. So that that's my thesis. That's why I did it the way that I did. I can say more about that, if you're interested.

GT 23:41 Definitely, keep going.

Steven 23:43 When Joseph intentionally plans to remember the vision, that is what's called in memory studies, Strategic Retrieval. He's going to write autobiography. He sits down. He's intentional about it. When that happens, a kind of brain work and a psychological response occurs, where he seeks for the beginning of the story, the middle of the story and the end of the story. When he does that, he is going to associate those points in time with emotion. He's going to associate past events with emotion. He's going to feel. He's going to have a psychological reaction. When Joseph intentionally thinks about writing autobiography, as he's doing in both 1832 and 1838, in other words, he's going to have an emotional response that he's going to write to.

Steven 24:47 So, my theory is that in 1832, his attempt is to minimize the tension between him and the culture. The minister represents the world Joseph lives in, the culture he has to inhabit. There's tension between that culture and Joseph Smith, the revelator, the bringer forth of the Book of Mormon. Joseph, I think, wants to minimize that tension just as you or I might want to minimize tensions between us and the main currents we have to swim in. So, I believe his way of doing that is to tell the 1832 story the way he does. I don't think he's satisfied with it, though. I don't think when he's done with it, it feels like it explains his present very well. He responds to his past in a way that might be psychologically healing, or he hopes will be minimally objectionable to the Minister and the whole culture the minister represents. But when Joseph is done, I don't think he feels like the 1832 memory is accurate to who he is in the summer or fall of 1832, the recipient of Doctrine & Covenants, Section 76, of Section 84.

GT 26:25 Section 76 occurred in 1832 as well, right?

Steven 26:27 February 1832. Then, late that summer, [section] 84 [is revealed.] So, you've got these expansive revelations that go well beyond evangelical soteriology. They go well beyond the Book of Mormon, in terms of the doctrines of salvation and priesthood work and mediation of priesthood, through ordinances and covenants and temple work. This is beginning to be unfolded in the exaltation doctrines in Section 76, and how that applies to temple work in section 84. So, Joseph, as a present Prophet, and President of The Church of Jesus Christ, in the summer and fall of 1832, is not the hurt kid, the teenager, who a few days after the vision, was rejected by a Methodist minister. The sort of the gap, the gulf between those selves that he has, factors into the way he remembers the vision. What I'm saying is, I think that his 1832 telling of the vision is a way to psychologically avoid the pain, the trauma of being rejected, the way he was when he first told the vision. But I believe, as he tells it that way, it's unsatisfying to him in his present role, in his grown-up role, his prophetic role. It doesn't sound like the President of the Church of Jesus Christ and the revelator of Section 76 and 84. Not everybody sees it that way. I understand that. This is a lot of theoretical, hypothetical work on my part. Ann Taves and others have articulated other positions and other ways of reading the same evidence. I'm open to all of those.

I'm excited to hear any kind of reading of the evidence that's responsible and engages it on a serious level. I'm not sure I'm right. But I was really, really compelled by Professor Allen's interpretive cue that that run-in with the Methodist minister was formative for Joseph, that it was traumatic, and that it shaped ever after the way Joseph responded. So, I'm saying that in 1832, he responds in one way.

GT 29:15 Is this a way where he's trying to act mature and trying to downplay the persecution? Is that what you're trying to tell me?

Steven 29:22 What I'm trying to say is that in 1832, he tells the vision in a way that is minimally conflicting with the evangelical Protestantism, [which] is the dominant cultural scene for him. In other words, it's in a way that's going to be acceptable to that Methodist minister, perhaps. Now, it's one way to read the evidence. You could make cases against it.

GT 29:59 Well, let me let me throw another case against it then, if I may.

Steven 30:02 Please do.

GT 30:03 Dan [Vogel] is the guy that I've talked to first about this. I'm going to probably bring a lot of Dan Vogel in here.

Steven 30:11 Great.

GT 30:11 One of Dan's issues with the 1838 account is he's noted some anachronisms. Dan believes that Joseph had a born-again experience. There was no vision, but he had a born-again experience in either 1820 or 1821. I mean, we're not super sure if it was 1820 or 1821: "It was my 14th year, my 15th year."

Steven 30:35 Yep.

GT 30:36 But somewhere around there, Joseph had a spiritual experience. So, Dan makes the case that--Methodists, especially, (so it's interesting to talk about the Methodist minister,) wouldn't have batted an eye because a lot of Methodists of the day had visions of Jesus.[4] So, it doesn't make sense that he came out in, let's say, 1820, and said, "I told my mother, Presbyterianism is false." Because Dan makes the case that when Joseph was 12 years old, he already knew all the churches were false. So, he sees that as an anachronism, and that that wouldn't have even happened. Then, the second thing is, and I know you get into this in part three, you talk about Wesley Walters back in I believe it was the 1960's or 70's. [Walters] says that there were no revivals in Palmyra until 1823, or 24. So, Dan makes the case that he couldn't have told anybody in 1820 that all the churches were false. There were no revivals. That didn't happen for three or four more years. So, what do you make of Dan's interpretation of those events?

Steven 31:57 Well, just like me, you have to read meaning into the facts. Wesley Walters, or Dan Vogel, or Ann Taves, or Steve Harper has to read meaning into the facts. All we know are a few facts. We often think we know more than we know, right? To decide whether something is anachronistic or not, how do you know? Do you have enough pieces of the puzzle to know if it's anachronistic?

GT 32:27 If you know there weren't revivals until 1823..

Steven 32:30 Do you know there weren't? No. You know that there's no evidence in the newspaper, for example. So, Wesley Walters takes the geographical area to be Palmyra village, and he shows that there are no newspaper accounts of camp meetings in the Palmyra village area in the 1820 window. That's what he knows. So let me be crystal clear. The fact is that he overstated it. Milton Backman did find a reference to a camp meeting in early 1820 in a Palmyra newspaper. So, Wesley Walters knows that the facts are, that in the records he researched, there was little to no mention of

---

[4] See our interview at https://gospeltangents.com/2019/06/methodist-visions/

unusual religious excitement in Palmyra village in 1820. Well, what he doesn't know is, is there unusual is excitement in the "whole district of country where we lived," right? That's Joseph's line. Joseph doesn't say Palmyra village. He says, "the whole district of country, indeed the whole region of country."

Steven 33:47 Joseph locates the unusual religious excitement around Manchester, which is actually where his family lives. They don't live in Palmyra, at the time of the vision or within a couple of years of it. So, you can't decide whether something's anachronistic or not, if you are deciding all the parameters of that. You can't be too close-minded about what Joseph means. One danger is not listening to Joseph well enough, deciding what he means. This is, I think, a problem with quite a lot of people, believers, unbelievers. They think they know what he means before they know what he means. So, I'm not sure I know everything he means, but I am more inclined to let him explain himself. I'm inclined to listen to him and trust him. I believe he tells an accurate story. Now, I'm not saying it's not distorted. I think he probably did blend memories about Presbyterianism. The idea of saying, "Mom, I know for myself Presbyterianism isn't true." I wouldn't be surprised if that's a later 1820s memory.

GT 35:05 Yeah. Because doesn't his mother and sister join the Presbyterians about 1823?

Steven 35:10 We don't know when they join. That's another thing people assume. We do not know when they joined. The records don't exist. We know when they leave the Western Presbyterian Church. We don't know when they join. If we did, it might help us sort through some of these things. Assuming that we know when they did is a problem.

GT 35:27 Well, aren't there Presbyterian newspapers around 1823-24?

Steven 35:31 Yeah.

GT 35:32 So does it seem likely that that's when that would have happened?

Steven 35:35 It does to me. In other words, it seems likely to me that, "Mom, I've learned for myself, [that] Presbyterianism isn't true," is likely an 1822, '23, '24 conversation. I wouldn't be surprised about that. But I don't take that to mean that--if Joseph blended that memory with the day he came home from the grove, and a later memory, I'd be okay with that. The reason I say that is because that line is a later emendation. That's a redaction to his manuscript history. It's not there in the original flow of the writing. Joseph didn't remember that, that earliest day.

GT 36:16 In his 1838 account?

Steven 36:19 That's an 1842 Willard Richards insertion, probably at Joseph's behest. Joseph might be saying, "Willard, I remember. Put that in, too." It wouldn't be a surprise to me if Joseph is blending memories. We all do this. This is not a conspiracy. He's not manipulating anything. If he's doing what I think he might be doing, he's doing it unconsciously. That is, he's remembering an event that happened later, as if it was part of the event that involves the grove and his first prayer.

GT 36:52 Well, that's what I said to Dan.[5] I said, "So what if he had a born-again experience in 1820, he had another experience in 1823, and just combined those together? And maybe the 1823 was when he saw somebody?" I mean, is that a possibility? It's not really in the historical record, right?

Steven 37:09 Yeah, it's all possible. I trust Joseph, though. I find no reason to disbelieve his accounts. In other words, Walters and Vogel, they read the accounts with a hermeneutic of suspicion. It's

---

[5] See our interview at https://gospeltangents.com/2019/06/first-vision-conflicts/

impossible to them that Joseph Smith was telling the truth about seeing God and Christ in the grove. It's not impossible to me, and that makes all the difference. I read Joseph with a hermeneutic of trust. Other people would approach him and say, "He can't possibly be telling the truth, so I must find an alternative explanation for why he says what he says." I read him and I say, "I think he is telling me the truth." So I take him pretty well at face value. Either of those are live possibilities. We all know the same evidence. I don't know more facts than they do, and they don't know more facts than I do about these historical events.

GT 38:10 I am curious, because you did say you found something from 1820 in Palmyra. Could you tell us a little bit more about that?

Steven 38:15 Yes, Professor Backman did. So, after Wesley Walters published his pathbreaking article where he said, "You can't prove whether a vision happened or didn't, but you can prove whether there was unusual religious excitement on the subject of religion, when and where Joseph said there was. I dug into all the newspaper records and all the church records at Palmyra Village at that time, and I found there was no evidence of unusual religious excitement." Therefore, there could have been no resulting revelation. It was not a catalyst for an early spring 1820 vision.

Steven: Well, that unleashed a whole bunch of research. What Professor Walters did, though ingenious, was fallible in a couple of ways. It argued an irrelevant proof in one sense, and it argued a negative proof in another. So, people set out to see what other evidence there might be and among these people was Milton Backman, a University of Pennsylvania Ph.D. [He's a] well-educated Latter-day Saint. He dug into the 'whole region of country.' He used Joseph Smith's geographical scope. Joseph used the Methodist term: the whole district of country seemed affected by the unusual religious excitement. So, where Wesley Walters casts his net small in Palmyra Village, a few miles north of Joseph Smith's farm, Professor Backman casts his net wide around the whole area of country, the whole district.

Steven 39:56 What he found was lots of evidence for unusual excitement on the subject of religion. The word revival comes up often, as if that's the measuring stick. A revival is the measuring stick. What often happens is people think a revival equals a camp meeting. All those things are related, but they're not all the same thing. If you confuse them for the same thing, you might mistake what you're looking at. So, there is evidence for a camp meeting in the newspaper in Palmyra in 1820. Professor Backman found it. He quoted in his resulting article and work. But is that an unusual excitement on the subject of religion? Professor Backman didn't think that one mention of that was, but he found plenty of examples of spikes in church attendance and church membership in various churches within a [radius of] 5, 10, 15-mile concentric circles. He found, in other words, evidence for unusual excitement on the subject of religion in the region or district of country that Joseph was saying.

Steven 41:10 He, [Professor Backman] also, didn't circumscribe it so much in time, as Reverend Walters did. Joseph didn't say it happened in the first days of 1820. Joseph gives more possible time for that unusual excitement. If you reach back into mid-1819, you find Methodists having conference meetings within a day's walk of Joseph's home, hundreds of Methodist ministers convening in this area. They'd have their conference meetings, and then they would spread out into the villages and preach. That happens in 1819. It happens again in 1820 within, again, a day's walk. It's not credible to argue that Joseph Smith could not have any basis for concluding that there was unusual excitement on the subject of religion in the district of country where he lived. That's simply hiding evidence. Now, how you interpret that evidence that's up to you, but to say it doesn't exist is irresponsible.

GT 42:15 Okay, so you think that there's evidence of those as early as 1820?

Steven 42:19 Absolutely.

GT 42:20 So, it's funny that we keep going from Methodist to Presbyterian.

Steven 42:24 They're both active. The Presbyterians are not too out there. They've got a couple of impressive churches in Palmyra Village, settled ministers, well-educated ministers. They are the folks who've been around longest. They have the market share. They are the most respectable. Lucy Mack Smith's heritage goes back into this tradition. She aspires to be a member of the Western Presbyterian Church. In the meantime, you've got these upstarts: Methodists. They're brand new, relatively speaking. They have ridden into town, literally on horseback--the famous Methodist preacher on horseback, that image from the 19th century that's going to become iconic. He's going to come into town. He doesn't even have a church. He just says, "Way down on the campground on the Geneva road, we're going to have a meeting. I'm going to exhort and anybody who needs to be saved can come there and we will get to work." This is appealing to lots and lots of people for all kinds of reasons. We know this is happening.

Steven 43:38 One reason we know this is happening is around the region and in the district of country where Joseph lives is the journals of a fellow name Benijah Williams. His journals are now at the University of California, Santa Barbara. They've got them in the library. Professor Taves has read them. I've read them. Others have read them. They evidence the kind of thing that a Methodist minister is doing and thinking about and worried about in that time and place. It's just what Joseph Smith is talking about. It's not credible to say that you can totally impeach him, that his facts are totally false or anachronistic. That's just a way of saying, "I select some facts and reject others and therefore find Joseph is anachronistic." I don't think that's a good way to do it.

Did Methodist Minister Scold Young Joseph?

Introduction

As we mentioned in a previous conversation with Dan Vogel, Joseph Smith's First Vision was quite similar to Methodist visions of Christ of the day.[6] Dan says it doesn't make sense for a Methodist minister to question Joseph's vision. I asked Dr. Steven Harper, author of "First Vision: Memory and Mormon Origins" to weigh in on this issue. Check out our conversation….

Interview

GT 44:41 I wonder about this other issue. Methodists were known for having visions, so, it seems a little strange that a Methodist minister would reject Joseph. Could it have been a Presbyterian minister, because they weren't as into ecstatic religious experiences, were they?

Steven 45:00 You're right about that, but it's probably a Methodist minister. I don't think Joseph is mistaken about that. Let me give you a potential interpretation of the facts that make sense. So, right now you're feeling like there's incongruity in Joseph's story. If he had reported a vision…

GT 45:19 I'm trying to give Dan's view.

Steven 45:20 Right. That's what he's saying, exactly. This is fun for me. I'm having a great time.

GT 45:26 This is good.

---

[6] See our interview at https://gospeltangents.com/2019/06/methodist-visions/

Steven 45:27 I hope you are, too.

GT 45:27 I hope Dan's watching.

Steven 45:28 Yeah, that would be fun. So, here are the facts. The facts are Methodists are active in the area. Joseph Smith says that he is partial to Methodism. He has some desire to be united with them. He tells Alexander Neibaur that he attends Methodist meeting and wants to feel and shout like the rest. He wants ecstatic religion. He wants to be overwhelmed by the Holy Spirit. A vision would be great. Right? [He wants] something that would confirm to him that the Presbyterian God of John Calvin, the angry God who abhors him and has already, in his sovereign will, destined him to hell, most likely, is not the real God.

GT 45:52 Wasn't it a Presbyterian that said Alvin was going to go to hell?

Steven 46:17 Yes, the Reverend Benjamin Stockton, exactly. You've got it.

GT 46:21 Okay, and Joseph Smith, Sr. wasn't very happy about it.

Steven 46:24 No, because it just doesn't sound very satisfying.

GT 46:27 Well, no wonder Joseph didn't like the Presbyterians.

Steven 46:29 Yeah, right. So, Joseph is inclined by heart to Methodism, where you can do something about it. You're still saved by the grace of God. You're still redeemed by Jesus Christ, but he endows you with the gift of His grace that enables you to come to him. Whereas, in the Presbyterian teaching of the time, God has made the decision of your salvation or damnation, by his arbitrary sovereign will, and there's nothing you can do about it. Joseph feels like, in his head, that's probably right, because I've been convicted of my sins and try as I might, nothing I do seems to matter.

Steven 47:12 But, in his heart, he holds on to the hope that the Methodists are right. He prefers that soteriology, that doctrine of salvation. So, this in mind, Joseph is not attuned to the fine points of debate, even inside the Methodist clergy. He doesn't know, as a later author put it, that orthodoxy became Methodized, and then Methodism became orthodox. In other words, he's not aware of what the Methodist ministers are aware of. That means that he thinks that going into the woods and having a vision is evidence of a Methodist conversion. It finally worked. The Methodists told me that might work. It was a Methodist minister, who said, "If you lack wisdom, ask God. I did everything they said, and I tried it and tried it before and it never worked, and, finally, it worked." So, Joseph's initial interpretation of his experience is, "I have now a Methodist conversion." What you do in that case, is you report it to the Minister. You get validated. He's shocked when he gets anything but validated, and so that's the point, right? You're saying, "Well, wouldn't a Methodist minister say, 'Yeah, that was a great vision you had.'" Not necessarily.

Steven 48:38 Right. Think about reasons why that might not be the default response. This Methodist minister may be aware that Methodism is trending toward enthusiasm, which is not a positive term in those days. That means to be crazy, or it's beginning to be [thought of as being crazy.]" There are some in the Methodist ministry, who are trying to pull back from that over-enthusiastic response. Lorenzo Dow is still going, and he's still working people into frenzy, but some of the Methodist clergy are saying, "Oh, that's just a little too weird for me." It's also the case that John Wesley, the founder of Methodism, has prophesied, "Look. We're going to grow like gangbusters, and the risk we run is becoming formalists." We might grow exponentially and get to a point where we're like everybody else where we speak of God with our mouths, but we deny the power thereof. We have a form of godliness, but we deny His power. "Don't ever do that," John Wesley says. So let's say you're a Methodist minister, and you've been influential in getting this feeling among the people that they can come to Christ. It's all good until you see maybe some people getting a little excessive for your comfort level, maybe going a little too far. Then one of them comes to you and says, "Guess what? It worked. I saw God and Christ in the woods, and guess what they said? Everybody here, including you, sir, have a form of godliness, but you deny the power, thereof." That's the cue for the Methodist minister to say, "[No.]"

GT 48:39 You think that's what Joseph Smith said to a Methodist?

Steven 50:32 I don't know. Nobody knows, but it's possible. You see, what I'm saying here is there are alternative ways to interpret the facts. So, the way that has Joseph Smith being unreliable, that's one interpretation. But it's not the only one. It's not the only credible one. It's not the only one you can make if you're smart, and the other ones are dumb. There are believing ways to understand Joseph that fit perfectly well with his time and place. He doesn't have to be impeached as anachronistic or unreliable as a rememberer. And nobody can verify it. It can be offered as an alternative. Nobody can verify my interpretation either. So, all we can do is read the facts and interpret them in the best way we can with the most context. We either do that with a believing bias, or an unbelieving bias, neither of which make us necessarily more intelligent or better at interpreting the facts. So, what I want people to understand is that you don't have to default to my interpretation, or Ann Taves, Dan Vogel's or whatever. Read this evidence, yourself, and come to your own conclusions. Don't let people who think we're smart like myself, make you feel like you're dumb. You're smart enough to make your own judgments about these things. Learn the facts. The facts are more easily accessible than they've ever been before. Listen to what Joseph says. Read all his accounts. Learn as much as you can about the context. Don't take Wesley Walters word for the context, or Milton Backman just by default. Read it all.

GT 52:22 Is there a place where you could you can actually [read all of the accounts?] I was just thinking of this yesterday. Somebody should write a book where you've got the 1832, 1835, 1838, 1842 accounts, so you can kind of look at them side by side.

Steven 52:39 I've done that.

GT 52:39 You've done that?

Steven 52:40 Yeah, I wrote a little book for Deseret [Book] called, "Joseph Smith's First Vision."[7] It's a seeker's guide to the accounts.

GT 52:45 It actually has the transcript of what he said?

Steven 52:49 Yeah, all of them are transcribed. A better place to go, though, is the Joseph Smith Papers website. They've got the images there.[8] They've got the transcriptions there, and James Allen's article, really, he was a pioneer. It was in 1970, or 71, when he published in the Improvement Era, an article called, *Eight Accounts of Joseph's Vision: What We Can Learn From Them?*[9] I think that's the title or something close [to that].

GT 53:18 Yeah, because there were four that were essentially either written or commissioned by Joseph, and then we have...

Steven 53:25 Five, now, that we know of...

GT 53:27 Five?

Steven 53:27 Five that were contemporary.

GT 53:29 So, we've got 1832, 1835, 1838, 1842...

Steven 53:32 Yep.

---

[7] Can be purchased at https://amzn.to/3xfx245
[8] See https://www.josephsmithpapers.org/site/accounts-of-the-first-vision
[9] See https://archive.bookofmormoncentral.org/content/eight-contemporary-accounts-joseph-smiths-first-vision-what-do-we-learn-them

GT 53:33 What am I missing?

Steven 53:34 We've got Orson Pratt, so, five contemporary accounts. You've got the four primary accounts.

GT 53:40 Those are the four primary, okay.

Steven 53:42 And then Orson Pratt, Orson Hyde, Levi Richards, David Wight, and Alexander Neibaur all created accounts of the vision that they heard Joseph give during his lifetime.

GT 53:56 Okay, so there are basically nine contemporary accounts.

Steven 53:58 So far, and it depends [how you count.] There's a 14 November 1835 entry in Joseph's journal where he says, "I told the Erastas Holmes about my vision of angels." Some people say there's an account of the vision. Some people say no, let's just him telling he gave an account of the vision, but it doesn't actually record any details. So, you can hear different people say there's five primary accounts or four. They'll do the same thing with the secondary ones, but generally speaking, the *Joseph Smith Papers*, for example, said there are nine accounts: four primary ones, five secondary ones, and here they are.

GT 54:36 I'll have to check that out.

Comparing the Primary Accounts

Introduction

What are the main differences between the First Vision accounts? Why are they different, and are these differences significant? Dr. Steven Harper is the author of "First Vision: Memory and Mormon Origins" and he will weigh in on these issues. Check out our conversation....

Interview

GT 54:37 So let's talk a little bit about the 1835 account. How is that different than both 1832 and 1838?

Steven 54:44 Excellent question. So, the 1832 and 38 are autobiographies. They are strategic memories. Joseph has stress and anxiety associated with strategic retrieval of his memory that he doesn't have when it's a spontaneous memory. So, the 1835 telling is a spontaneous retrieval. Joseph is not planning to write anything. He's not planning to tell the story of his First Vision. He's talking to this fellow from the east, Robert Matthews, and they start comparing prophetic credentials. This guy thinks he's a great spiritual leader. He's heard Joseph is, so, he's come to see him, kind of to compare notes. Maybe, there's kind of a subtle competition going on between them. I think, at least Matthews is trying to figure out if he might ally himself with Joseph Smith in some way or other.

Steven 55:37 So, they're very curious about each other and they want to know what's going on inside each other's brains. They start swapping credentials for what makes them a prophet. Joseph says, "Well, let me tell you how the Book of Mormon came forth. The first thing that happened is, I was worried about matters that involve eternal consequences, and I worried about it a lot. I had great anxiety. I was distressed and perplexed, and I went to the woods to pray. I saw a fire, and then one personage revealed another. It filled me with joy unspeakable [joy.]" It's a fast moving, relatively easy flow for Joseph. When you compare it to the autobiographies, you notice that it's not freighted with the concern about writing. The first thing Joseph does in both of his autobiographies is he offers a disclaimer about why he can't write well.

GT 56:35 So, the 1835 is not written by Joseph.

Steven 56:38 That's right, it's written by Warren Parrish. Parrish captures it.

GT 56:40 Oh, Warren Parrish.

Steven 56:40 Parrish captures it and puts it into his journal. Joseph is not writing it. He's not thinking about writing it. He's not thinking about, "What's the beginning of the story, the middle of story, the end of the story. How do I structure this narrative?" He's just spilling it out. It comes naturally to him, in that sense. It's much easier work for him when he tells it like that, than it is when he writes it. We now know that he tells it like that quite a bit in this middle 1830s period, much more than we used to think. He's telling it that way by shortly after, if not at the same time or before he writes the 1832 autobiography. So, 1835 memory is really cool. I think one of the most telling things about it is, it doesn't seem to cause Joseph Smith the psychological need to reconcile with or deal with that Methodist minister's rejection. It's one of the things I argue in the book is the 1832 memory is an effort to make good with or at least not offend the minister or the whole world the minister represents, and that Joseph isn't very satisfied with his memory as a result of that effort. Then, I argue that the 1838 memory is an effort to take that minister head on. This is Joseph in the worst year of his life. He is in a persecution mindset. Notice how many times that account says hot persecution, the bitterest persecution.

GT 56:44 I know Dan Vogel mentions that.[10]

---

[10] See https://gospeltangents.com/2019/06/first-vision-conflicts/ and https://gospeltangents.com/2019/06/why-pious-fraud-ticks-off-everyone/

Steven 58:22 It is definitely the present that gives us that version of the past. It's saturated with persecution. In that mode, Joseph Smith spits venom at the clergy. He calls the Methodists "priests" three times. He knows that that's a way to offend.

GT 58:46 Because he's comparing [the Methodist minister] to the Catholic priest. Is that right?

Steven 58:49 Right. The people who are not liked very much, like the Latter-day Saints in Protestant America in the 1830s, are Catholics. So, if you want to call a Methodist clergyman a name, you call him a priest and compare him to the hireling priest. You label him with those mean names that people call Catholics. I think Joseph knows he's doing that when he does it. And from the first line, "Owing to the many reports put in circulation by evil disposed and designing people, I'm going to tell you the truth." Then Joseph tells the story, and then notice that interpretive memory, the last fact he gives us, the last piece of factual memory is, "A few days after the vision. I was in company with the Methodist minister. I told him, and he rejected me." Then Joseph launches into this passage of interpretive memory about how it felt, and this is how it feels in 1838 to reflect on this whole history. When Joseph searches his past for the origin of persecution, it started with that rejection. And he interprets it. He internalizes it. It's heavy. It's traumatic. It's emotional. That feeling that comes out of that 1838 memory is because of all those combining factors, the original rejection, all of the experience since, the persecution of the present, and Joseph remembering, strategically in that context, he's pretty upset. You can feel it as you read that.

GT 1:00:33 Interesting. Tell us more about the 1842 [account.] I know Dan kind of collapsed 1838 and 1842, because he said they really weren't that different. Are there any differences there?

Steven 1:00:45 There's not a whole lot [differences] after the '38 [account], which is intensely interesting for all kinds of interesting reasons. The '42 [account] is sort of ho hum. It's a response to John Wentworth's invitation for Joseph to provide his own story. So, Joseph welcomes it. This present is a whole lot more peaceful than the 1838 present. The 1838 present that gives us the past, as Joseph remembered it, is full of defensiveness to persecution and a response to that minister, and all the opposition Joseph has faced ever since. Well, when it's few years removed, and Missouri is in the rearview mirror, and Carthage isn't on the horizon, and you get a nice note from a prominent Chicago newspaper guy who says, "Hey, you've got some notoriety and a buddy of mine is going to write a history of New Hampshire and you spent part of your growing up years there. Why don't you tell us your own story?" That elicits from Joseph, a very calmed down, PR-based, PR-minded version of the story, with a little Latin thrown into it. This is Joseph trying to be his most refined self. In this version of the story, God doesn't say all their creeds are an abomination. He says, as Joseph paraphrases it. "He told me that all the churches were believing in incorrect doctrines." Same story, but with a different spirit.

GT: He toned it down a lot.

Steven: It has a different tone to it. I don't know that Joseph did that on purpose. It could be just the way we respond in a present. A much more pleasant present, gives us a less embittered past.

GT 1:02:36 So, in 1838, let's talk about what other things were going on. That's when he got kicked out of Kirtland. Is that right?

Steven 1:02:45 Yes.

GT 1:02:45 The bank collapsed.[11]

---

[11] See our interviews with Dr. Mark Staker at https://gospeltangents.com/category/kirtland-banking-crisis/

Steven 1:02:46 They started that account after he's driven from Ohio. He leaves Ohio in the dead of the night, flees out of there. He's quite worried about how things are going to go and arrives in Missouri, just in time for the bottom to fall out there. So he starts that...

GT 1:03:04 Hawn's Mill was that same year, as well, wasn't it?[12]

Steven 1:03:06 Yes, that fall, exactly.

GT 1:03:08 That was quite a year. You've got Hawn's mill at the end.

Steven 1:03:10 That was the worst year of his life.

GT 1:03:11 You've got the bank crisis at the beginning.

Steven 1:03:13 What we know is that Joseph started that 1838 history in 1838. He starts before everything goes south in Missouri, but the document we have is an 1839 document, even though it has an internal date that says 1838. It's an 1839 draft. What we know is that between the first draft and the draft we have was the worst year of Joseph's life: Liberty Jail, Hawn's Mill, the siege at Far West, everything awful, the Extermination Order.[13] So, that's the taste in Joseph's mouth when he puts that document together. It's hard to tell exactly what comes from 1838, and what comes from 1839. But we can say with good accuracy that the context of the whole thing is bitter for Joseph.

GT 1:04:13 So that explains why he says all the churches are an abomination. He's using a lot of really emotional type terms.

---

[12] See our interviews with Dr. Alex Baugh at https://gospeltangents.com/category/hawns-mill-massacre/

[13] See our interview at https://gospeltangents.com/2019/10/extermination-order-license-kill/

Steven 1:04:19 I'm not positive that explains it completely. I'm not sure that the Lord doesn't say, "All their creeds are an abomination." He may very well have said that. I would have said it if I were him, for good reason. So, I don't know. There's no way for me to know for sure what exactly the Lord says in the vision. But I'm saying that I think a sound way to interpret the tone of the 1838 document is to see it in the present context of the worst year of Joseph's life. That may give us the sort of defensive, defiant, "all their creeds are an abomination" kind of tone that we have.

GT 1:05:06 And then, by 1842, things are semi-calm in Nauvoo. He hadn't really started doing polygamy yet.

Steven 1:05:14 He's started that by then.

GT 1:05:15 He has, but it's secret at the time.

Steven 1:05:17 Right. Another way to see what I'm saying here is that we now have what we call a fair copy of the 1838-39 document. This is a Howard Coray draft of the same document. It comes from around 1840-41. So, again, it comes from a period when Missouri is fading into the background, and Carthage is not on the horizon yet. It's a peaceful lull. Nauvoo is a place where we've got a charter that protects us. We've been welcomed into the state. It's just such a different feeling from Missouri. In that present, Joseph cuts off the defensive beginning of the document. It just says, "I was born in Sharon," instead of "owing to the many reports put in circulation." Then, interestingly, too, that big interpretive memory at the end, after he says, "I was rejected by the Methodist minister, and it felt like I was Paul before Agrippa, and I knew it and I knew God knew it." That part's all gone, too, and it's just the factual memory, almost the exact same factual memory. But the interpretive parts, the parts that reflect the defensive present of 1838-39, he says, "Those don't feel right." They don't feel quite the same in the 1840-41 present. I think that's the reason why those bookends are gone. So it's a useful way to see this point about memory and how memory works. We make memories out of our present, and our past. The feeling of the present has a lot to do with the tone, the spirit of the past that we recall. It's not that the facts are wrong, necessarily, but the way the facts are presented and interpreted and the meaning that you breathe into them, even with the vocabulary, is very much dependent on the mood we're in when we remember.

Orson Pratt's First Vision Influence

Introduction

Early Mormon apostle Orson Pratt probably did more to keep the memories of the First Vision alive in the LDS Church than any other person. In our next conversation with Dr. Steven Harper, author of "<u>First Vision: Memory and Mormon Origins</u>" we'll talk about Orson's outsized influence. We'll also talk about how some modern critics view the First Vision. Check out our conversation....

Interview

GT 1:07:25 Okay, so let's wrap this up with a few other things. I just wanted to talk a little bit about Orson Pratt and how he influenced institutional memory. Can you talk about him especially, and George A. Smith, and anybody else you want to talk about?

Steven 1:07:40 I think it's likely that if not for Orson Pratt, we would have a much diminished collective memory, as Latter-day Saints of Joseph Smith's First Vision. He is the foremost selector and relator and repeater of the vision, to use the technical terms that Thomas Anastasio and his colleagues use, for the people who choose what we remember. That happens because someone selects it, someone repeats it often, and relates it to other important components of our shared story. Nobody did that like Orson Pratt did that in the middle of the 19th century. He got the story from Joseph Smith's own mouth.

Steven 1:07:42 Orson Pratt heard Joseph tell his First Vision on his way to Scotland on his mission. Joseph was on his mission to Washington, D.C. to seek redress for Missouri grievances. He and Orson Pratt cross paths in the Delaware River Valley. Orson learns the story from Joseph. He writes it in a missionary pamphlet in

Scotland. That circulates all over the globe. Orson, ever after, tells that story. He tells the First Vision often. He tells it early. He coined the term "First Vision," as far as we can tell. It's in his writing in 1849, that those two words are used together for the first time in the historical record. Throughout the mid-decades of the 19th century, other church leaders are not telling the vision nearly as often, and they're not telling it in the same way. Even though Joseph Smith's records now and they've been published in the Church newspaper, Joseph Smith's History will be published in the Pearl of Great Price in Britain in 1851. It'll be canonized in 1880. But in that 30 years, you find quite remarkable variations on the story from George A. Smith, John Taylor, Brigham Young, and others. So, it's Orson Pratt, who tells the story pretty much the way Joseph tells it and repeats it and keeps it on the forefront of minds. Finally, then, it gets canonized. We remember it the way we remember it today, largely because of the work that Orson Pratt did.

GT 1:10:15 Yeah, he was the Church Historian from...

Steven 1:10:19 For part of that time in the 1870's.

GT 1:10:21 He's also the one who first published the Pearl of Great Price, which included it.

Steven 1:10:26 That's Franklin Richards.

GT 1:10:28 Oh, that was Franklin Richards, okay.

Steven 1:10:29 Franklin Richards, in 1851, as president the British mission, selected the contents of the Pearl of Great Price.

GT 1:10:39 I conflated my memory there. (Chuckling)

Steven 1:10:42 You and me and everybody else.

GT 1:10:45 Alright, so, we can really thank Orson for kind of a modern telling.

Steven 1:10:52 We can.

GT 1:10:52 Because without him, it might not still be as well-known. Is that true?

Steven 1:10:57 I think that's likely the case, yes. It's Milton Backman, really, who figured that out. I don't want to pretend to own that research. It's Milton Backman who realized, from his research, that Orson Pratt played that key role. I built on Professor Backman's research, like I built on Professor Allen's and others, but he was the one who originated that argument.

GT 1:11:26 I know in the last part [of your book], we've talked a little bit about Wesley Walters. I know what caught my attention last night as I was leafing through the book, is you had mentioned there was a podcast with John Dehlin on *Mormon Stories* and it had Sandra Tanner [on it.] I guess you watched. I actually haven't seen it. I'll have to go back and watch it. What are the things that you think they got wrong, or you take issue with the most as they tell the stories of discrepancies with the First Vision?

Steven 1:11:57 Well, the fundamental thing is simply a different choice about how to interpret Joseph. Sandra Tanner knows her facts. She's actually pretty judicious about her facts. She and her late husband were pretty careful over time. They didn't exaggerate facts or try to sugarcoat things or hide things. So, I admire that about her. But the fact of the matter is, she chooses a hermeneutic of suspicion. She's skeptical of Joseph Smith. He can't be telling a true story if the version of Christianity that she has chosen is true. It's the same with Wesley Walters. So, the difference between my choices and hers are simply different versions of Christianity. That means that we have different views about whether we can trust Joseph Smith or not. So, we read the same documents. We know the same facts. John Dehlin, interestingly, has sort of migrated. He started out with a hermeneutic of trust in Joseph Smith. Then, over time he's migrated to a position of extreme suspicion. He no longer believes that Joseph can be trusted in his First Vision accounts. That's really the reason that he and Sandra Tanner were eye to eye on that particular episode of *Mormon Stories*, and probably why he wanted to hear her voice on that anyway and feature her because she represented that hermeneutic of suspicion. I don't know what he's believing at this point about Christ or whatever else, but he wanted to feature that distrustful or skeptical interpretation of Joseph.

GT 1:13:53 So, that's your biggest issue: you're a believer. They're not. We're going to look at facts differently, just based on our point of view. Is that right?

Steven 1:14:03 Yeah. I wouldn't call it an issue. That's just the way it is. The question might be asked, "Well, why do to people who know the same facts...

GT 1:14:15 I get that a lot.

Steven 1:14:15 ...and study the same historical records come to such dramatically different conclusions? It's because historians aren't endowed with some godlike capability of knowing. They only know the same facts that anyone else can know. Then, they just interpret the facts. Their interpretations are necessarily dependent on their biases and prejudices and choices, faith commitments, or lack thereof. Some people are under the impression that it's the facts of the matter that turn the tide. No, it isn't. The editors of the *Joseph Smith Papers* are believers. They know all the facts. Dan Vogel, Sandra Tanner, they know the facts. Everybody invested in this knows the facts. [We are all] reading the same documents and the same evidence. I've had really wonderful exchanges with Ann Taves, who knows the facts well. She studies them really carefully and arrives at different interpretations than I do. It's not that one of us knows the evidence better than the other. It's that we just make different choices about what the evidence means.

GT 1:15:12 Well, great. Is there anything we've missed in your book that you'd like to share?

Steven 1:15:36 No, I really appreciate you paying attention to it. It's gratifying to me. I sometimes wonder if anybody will care. My mom bought a copy and I think maybe a few other people have.

GT 1:15:48 I bought a copy.

Steven 1:15:49 Perfect. I've been waiting for it to hit the *New York Times* bestseller list, and so far, it hasn't climbed the list.

GT 1:15:58 (Chuckling)

Steven 1:15:58 It's gratifying to have your interest and attention to it. I appreciate it.

GT 1:16:03 Great. Well, Steven Harper, thank you so much for being here on *Gospel Tangents*. It's really appreciate it.

Steven 1:16:08 You bet. Thank you.

Additional Resources:

Check out our other interviews on the First Vision with Dan Vogel.

Dan Vogel on First Vision

Historian Dan Vogel is author of many books on Joseph Smith, including "Making of a Prophet."[14]

292: First Vision Conflicts
https://gospeltangents.com/2019/06/first-vision-conflicts/

291: 1835 Account of First Vision
https://gospeltangents.com/2019/06/1835-first-vision/

290: Making a Case for Melchizedek Priesthood in 1831?
https://gospeltangents.com/2019/06/melchizedek-priesthood-1831/

289: Methodist Visions
https://gospeltangents.com/2019/06/methodist-visions/

288: Why "Pious Fraud" Ticks off Everyone
https://gospeltangents.com/2019/06/why-pious-fraud-ticks-off-everyone/

287: Dan Vogel Was a McConkie Mormon!
https://gospeltangents.com/2019/06/dan-vogel-was-a-mcconkie-mormon-part-1/

---

[14] Can be purchased at https://amzn.to/3wLDFec

Last Thoughts

You can get our transcripts at our amazon.com author page. I've got a link here, but just do a search for Gospel Tangents interview, and you should be able to find a bunch of them there. Please subscribe at Patreon.com/gospeltangents. For $5 a month, you can hear the entire interview uncut and for $10 you can get a pdf copy. We've also got a $15 tier where if you want a physical copy, I'll be the first to send it to you, so please subscribe at Patreon or on our website at Gospeltangents.com. For our latest updates, please like our page at facebook.com/Gospeltangents and also check our twitter updates Gospel tangents. Please subscribe on our apple podcast page tinyurl.com/GospelTangents, or you can subscribe on your android device. Just do a search for Gospel Tangents. Thanks again for listening. Click here to subscribe, here for transcript and over here we've got some more of our great videos. Thanks again.

www.ingramcontent.com/pod-product-compliance
Lightning Source LLC
Chambersburg PA
CBHW050319220526
45465CB00005B/2045